Stochastic Computing

Stochastic Computing

John Sartori and Rakesh Kumar

University of Illinois at Urbana-Champaign
Urbana, IL 61801
USA
{sartori2@illinois.edu;rakeshk@illinois.edu}

Boston – Delft

Foundations and Trends® in Electronic Design Automation

Published, sold and distributed by:
now Publishers Inc.
PO Box 1024
Hanover, MA 02339
USA
Tel. +1-781-985-4510
www.nowpublishers.com
sales@nowpublishers.com

Outside North America:
now Publishers Inc.
PO Box 179
2600 AD Delft
The Netherlands
Tel. +31-6-51115274

The preferred citation for this publication is J. Sartori and R. Kumar, Stochastic Computing, Foundations and Trends® in Electronic Design Automation, vol 5, no 3, pp 153–210, 2011

ISBN: 978-1-60198-508-8
© 2011 J. Sartori and R. Kumar

All rights reserved. No part of this publication may be reproduced, stored in a retrieval system, or transmitted in any form or by any means, mechanical, photocopying, recording or otherwise, without prior written permission of the publishers.

Photocopying. In the USA: This journal is registered at the Copyright Clearance Center, Inc., 222 Rosewood Drive, Danvers, MA 01923. Authorization to photocopy items for internal or personal use, or the internal or personal use of specific clients, is granted by now Publishers Inc for users registered with the Copyright Clearance Center (CCC). The 'services' for users can be found on the internet at: www.copyright.com

For those organizations that have been granted a photocopy license, a separate system of payment has been arranged. Authorization does not extend to other kinds of copying, such as that for general distribution, for advertising or promotional purposes, for creating new collective works, or for resale. In the rest of the world: Permission to photocopy must be obtained from the copyright owner. Please apply to now Publishers Inc., PO Box 1024, Hanover, MA 02339, USA; Tel. +1-781-871-0245; www.nowpublishers.com; sales@nowpublishers.com

now Publishers Inc. has an exclusive license to publish this material worldwide. Permission to use this content must be obtained from the copyright license holder. Please apply to now Publishers, PO Box 179, 2600 AD Delft, The Netherlands, www.nowpublishers.com; e-mail: sales@nowpublishers.com

Foundations and Trends® in Electronic Design Automation

Volume 5 Issue 3, 2011

Editorial Board

Editor-in-Chief:
Sharad Malik
Department of Electrical Engineering
Princeton University
Princeton, NJ 08544

Editors

Robert K. Brayton (UC Berkeley)
Raul Camposano (Synopsys)
K.T. Tim Cheng (UC Santa Barbara)
Jason Cong (UCLA)
Masahiro Fujita (University of Tokyo)
Georges Gielen (KU Leuven)
Tom Henzinger (EPFL)
Andrew Kahng (UC San Diego)
Andreas Kuehlmann (Cadence Berkeley Labs)
Ralph Otten (TU Eindhoven)
Joel Phillips (Cadence Berkeley Labs)
Jonathan Rose (University of Toronto)
Rob Rutenbar (CMU)
Alberto Sangiovanni-Vincentelli (UC Berkeley)
Leon Stok (IBM Research)

Editorial Scope

Foundations and Trends® in Electronic Design Automation will publish survey and tutorial articles in the following topics:

- System Level Design
- Behavioral Synthesis
- Logic Design
- Verification
- Test

- Physical Design
- Circuit Level Design
- Reconfigurable Systems
- Analog Design

Information for Librarians

Foundations and Trends® in Electronic Design Automation, 2011, Volume 5, 4 issues. ISSN paper version 1551-3939. ISSN online version 1551-3947. Also available as a combined paper and online subscription.

Foundations and Trends® in
Electronic Design Automation
Vol. 5, No. 3 (2011) 153–210
© 2011 J. Sartori and R. Kumar
DOI: 10.1561/1000000021

Stochastic Computing

John Sartori and Rakesh Kumar

University of Illinois at Urbana-Champaign, 1308 W. Main St., Urbana, IL 61801, USA, {sartori2@illinois.edu;rakeshk@illinois.edu}

Abstract

As device sizes shrink, manufacturing challenges at the device level are resulting in increased variability in physical circuit characteristics. Exponentially increasing circuit density has not only brought about concerns in the reliable manufacturing of circuits but also has exaggerated variations in dynamic circuit behavior. The resulting uncertainty in performance, power, and reliability imposed by compounding static and dynamic nondeterminism threatens the continuation of Moore's law, which has been arguably the primary driving force behind technology and innovation for decades. This situation is exacerbated by emerging computing applications, which exert considerable power and performance pressure on processors. Paradoxically, the problem is not nondeterminism, *per se*, but rather the approaches that designers have used to deal with it. The traditional response to variability has been to enforce determinism on an increasingly nondeterministic substrate through guardbands. As variability in circuit behavior increases, achieving deterministic behavior becomes increasingly expensive, as performance and energy penalties must be paid to ensure that all devices work correctly under all possible conditions. As such, the benefits of technology scaling are vanishing, due to the overheads of dealing with hardware variations through traditional means. Clearly, *status quo* cannot continue.

Despite the above trends, the contract between hardware and software has, for the most part, remained unchanged. Software expects flawless results from hardware under all possible operating conditions. This rigid contract leaves potential performance gains and energy savings on the table, sacrificing efficiency in the common case in exchange for guaranteed correctness in all cases. However, as the marginal benefits of technology scaling continue to languish, a new vision for computing has begun to emerge. Rather than hiding variations under expensive guardbands, designers have begun to relax traditional correctness constraints and deliberately expose hardware variability to higher levels of the compute stack, thus tapping into potentially significant performance and energy benefits and also opening the potential for errors. Rather than paying the increasing price of hiding the true, stochastic nature of hardware, emerging *stochastic computing* techniques account for the inevitable variability and exploit it to increase efficiency. Stochastic computing techniques have been proposed at nearly all levels of the computing stack, including stochastic design optimizations, architecture frameworks, compiler optimizations, application transformations, programming language support, and testing techniques. In this monograph, we review work in the area of stochastic computing and discuss the promise and challenges of the field.

Contents

1 Introduction   1

2 Design-Level Techniques for Stochastic Computing   5

2.1 EVAL and BlueShift   8
2.2 Dynatune   9
2.3 CRISTA   10
2.4 Better-Than-Worst-Case Synthesis   11
2.5 Recovery-Driven Design   12
2.6 Gradual Slack Design   17

3 Architecture Frameworks for Stochastic Computing   21

3.1 Error-resilient System Architecture (ERSA)   21
3.2 Algorithmic Noise Tolerance (ANT)   22
3.3 Stochastic Sensor Network on Chip   23
3.4 Variation-Adaptive Stochastic Computer
     Organization (VASCO)   25
3.5 Architectural Principles for Stochastic Processors   26

4 Compiler Optimizations for Stochastic Computing   35

4.1 Critical Path Avoidance   36
4.2 Activity Throttling   37
4.3 Overlapping Errors   37

4.4   Recovery Cost-Aware Scheduling   37

4.5   Optimizing the Binary to Reduce Error Rate   38

5   Scheduling and Mapping for Stochastic Processors   45

**6   Application and Programming Language Support
for Stochastic Processors   47**

7   Testing Techniques   51

8   Other Work   55

9   Summary and Future Work   59

References   61

1

Introduction

The primary driver for innovations in computer systems has been the phenomenal scalability of the semiconductor manufacturing process, governed by Moore's law, that has allowed us to literally print circuits and systems growing at exponential capacities for the last three decades. The resulting exponentially reducing cost per function has resulted in an unprecedented penetration of technology in homes and beyond, leading to profound impacts on society and quality of life.

Moore's law has come under threat, however, due to the resulting exponentially deteriorating effects of material properties on chip reliability and power. As transistors become smaller (the oxide in a 22 nm process is only five atomic layers thick, and gate length is only 42 atoms across), it is becoming increasingly expensive for the current design and manufacturing technology to keep transistors functioning deterministically, even under normal operating conditions. There are three primary sources of nondeterminism [23]. First, decreasing transistor sizes lead to different transistors being doped differently during the manufacturing process, causing them to have nondeterministic electrical characteristics [24]. Second, transistors have become smaller than the wavelength of the light used to pattern them (by 6×) [2]. This causes nondeterminism in the dimensions and characteristics of the manufactured transistors. Finally, the unprecedented increase in the power density of chips, coupled with time and context-dependent variation in temperature and utilization across the chip, cause voltage and timing variations in circuits [7]. These variations are dynamic and largely nondeterministic. The most immediate impact of such nondeterminism is decreased chip yields. A growing number of parts are thrown away since they do not meet timing and power-related specifications. A 5% yield loss on a 90 nm process today directly translates into a cost to the manufacturer that exceeds 2× the design cost for a typical cell-phone manufacturer [19], arguably one of the highest volume parts. Clearly the *status quo* cannot continue. Left unaddressed, the entire computing and information technology industry will soon face the prospect of parts that neither scale in capability nor cost. We must find a solution to the nondeterminism problem if semiconductor technology and industry are to remain a viable driver of science innovation and technology capabilities for the future.

Paradoxically, the problem is not nondeterminism, *per se*, but how computer system designers approach it. Chip components no longer behave like the precisely chiseled machines of the past; yet, the basic approach to designing and operating computing machines has remained unchanged. While there have been many swings in computing platform paradigms, such as from general-purpose to specialized, and from single-core to multi-core, the contract between hardware and software has remained unchanged. This contract guarantees that hardware will return correct values for every computation, under all conditions. In other words, we demand hardware to be overdesigned to meet the mindsets in computer systems and software design of the past. Guardbands imposed upon hardware result in increased cost [12], because getting the last bit of performance incurs too much area and power overhead, especially if performance is to be optimized for all possible computations. Conservative guardbands also leave enormous performance and energy potential untapped, since the software assumes lower performance than what a majority of instances of that platform may be capable of attaining most of the time.

As the marginal benefits of technology scaling continue to languish, a new vision for computing has begun to emerge. Rather than hiding variations under expensive guardbands, designers have begun to relax traditional correctness constraints and deliberately expose hardware variability to higher levels of the compute stack, thus tapping into potentially significant energy benefits, but also opening up the potential for errors. Rather than paying the increasing price of hiding the true, stochastic nature of hardware, emerging *stochastic computing* techniques exploit error resilience by exposing hardware errors and allowing hardware reliability to be traded for increased energy efficiency.

Truly, designers have begun to embrace stochasticity at many layers of the compute stack [37]. Resulting *stochastic architecture frameworks* [20, 29, 38, 42] are structured around stochastic computing techniques, which they exploit to enable energy-reliability tradeoffs and increase efficiency in the face of nondeterministic hardware. For example, the Variation-Adaptive Stochastic Computer Organization (VASCO) [38] dynamically adapts its own hardware reliability based on workload characteristics and environmental conditions to maximize the energy and performance benefits of exploiting error resilience.

Architecture frameworks such as VASCO rely on *microarchitecture and design-level techniques that manipulate the error distribution* of hardware to enable energy-reliability tradeoffs and increase the efficiency of operating in the stochastic domain. Design-level stochastic computing techniques [11, 16, 18, 25, 26, 27, 33, 43, 44] aim to make stochastic computing itself more efficient. For example, recovery-driven design [26] assumes a stochastic operating condition and the availability of hardware or software error resilience and asks how circuits can be designed to be more efficient when errors are allowed even in the nominal case. Microarchitectural techniques [35] attempt to reduce the number of errors that a processor produces for a given operating condition. Compiler optimizations [36, 37] can further improve the extent of benefits on programmable stochastic architectures by manipulating the activity distribution of a stochastic processor to reduce the error rate or cost of error recovery.

Programming language [13, 32] and application [37, 39] support for stochastic processors facilitates programming for stochastic processors and enables more programs to be executed on stochastic processors. Testing techniques [4] are also being re-evaluated so as to not require all parts of the chip to function flawlessly. The focus is especially on increasing chip yields in spite of permanent faults.

Going forward, as variability continues to grow and designers continue to abandon traditional variability-mitigation practices and embrace stochastic computing, stochastic design optimizations, architecture frameworks, compiler optimizations, application transformations, programming language support, and testing techniques will be essential to maximize the potential of stochastic computing to increase performance and energy efficiency.

In the first section of this chapter, we discuss design-level optimization techniques that aim to enable energy-reliability tradeoffs and improve the efficiency of error resilient designs.

2

Design-Level Techniques for Stochastic Computing

At the heart of stochastic computing is the reality that performing error-free computation is expensive. Conventional design methodologies sacrifice energy and performance in order to achieve near perfection. Stochastic computing proposes a new vision for energy and performance efficiency in which some errors are allowed, as long as they are corrected or tolerated by hardware, software, or the end user. The efficiency of stochastic computation depends on how many errors are produced and how much it costs to correct or tolerate them. In this section, we discuss design-level techniques that manipulate the error distribution of hardware to enable processors to make energy-reliability tradeoffs, ultimately making stochastic computing more efficient.

Due to the high cost of ensuring error-free compuation, many stochastic designs target a better-than-worst-case or average case operating point by relaxing worst case design constraints. One increasingly popular technique for relaxing correctness guarantees and targeting average case operation is timing speculation [3, 12, 15, 17]. Conventional design practices ensure that all circuit paths can meet timing, even under worst case process, voltage, and temperature (PVT) variations. As a result, conventional designs are considerably over-designed for the average case. Timing speculative design techniques recognize the substantial power and performance overheads imposed by worst case design practices and relax them by scaling up the operating frequency (frequency overscaling) or scaling down the operating voltage (voltage overscaling). An overscaled design has higher performance or lower power than its counterpart worst case design. However, since the delay of some paths may now be greater than the clock period under certain conditions, timing violations can occur — where the correct output of a logic path has not reached the path output in time to be captured in the output register. To account for this occurrence, timing speculative designs often include mechanisms to prevent [6, 14, 28] or detect and correct [3, 12, 15, 17] errors. Since allowing, tolerating, or correcting errors costs performance, power, or output quality, the benefits achieved by timing speculative designs depend on the error rate that overscaling induces. First, we describe the factors that influence the timing error rate of a design. Then, we go on to describe stochastic design optimizations that aim to improve the efficiency of timing speculation by manipulating the error rate.

Many of the proposed techniques use voltage or frequency overscaling as a proxy for all variation-induced errors. Nonetheless, their conclusions should be applicable for other sources of timing variation as well.

The extent of energy benefits provided by timing speculation depends on the error rate of the timing speculative design. For example, in the context of voltage overscaling, benefits depend on how the error rate changes as voltage decreases. If the error rate increases steeply, only meager benefits are possible [25] due to high error recovery overheads or limited scalability. If the error rate increases gradually, greater benefits are possible.

While energy benefits depend on the timing error rate of the processor, the error rate itself depends on the timing slack and activity of the paths of the processor in the context of overscaling. Figure 2.1 shows an example slack distribution. The slack distribution is a histogram that shows the number of paths in a circuit that have a certain amount of timing slack (*clock period — path delay*). As voltage is scaled down, path delay increases, and path slack decreases. Likewise, as frequency

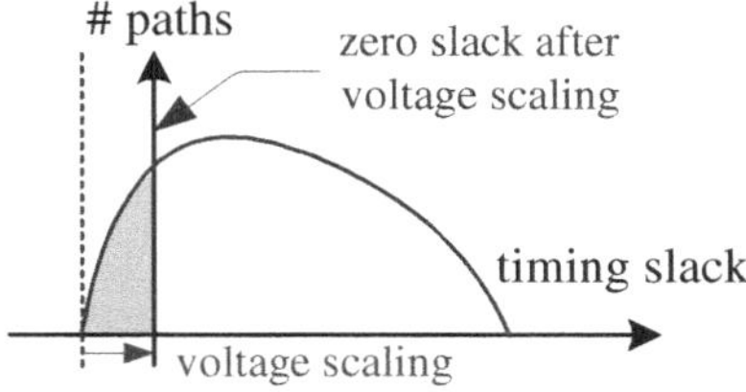

Fig. 2.1 Voltage or frequency scaling shifts the point of critical slack. Paths in the shaded region have negative slack and cause errors when toggled.

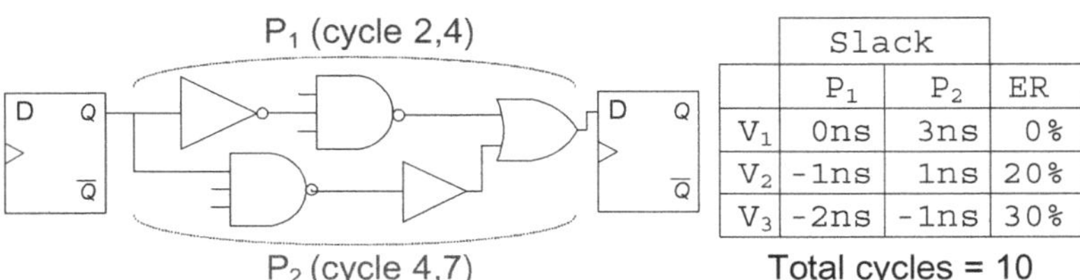

	Slack		
	P_1	P_2	ER
V_1	0ns	3ns	0%
V_2	-1ns	1ns	20%
V_3	-2ns	-1ns	30%

Total cycles = 10

Fig. 2.2 Slack and activity distributions determine the error rate. Error rate increases with overscaling, since more paths cause errors.

is scaled up, the clock period gets shorter, and path slack decreases. The slack distribution shows how many paths can potentially cause errors because they have negative slack (shaded region). Negative slack means that path delay is longer than the clock period.

From the slack distribution, it is clear which paths can cause errors at a given voltage or frequency. In order to determine the error rate of a processor, the activity of the negative slack paths must be known. A negative slack path causes a timing error when it toggles. Therefore, knowing the cycles in which any negative slack path toggles reveals the number of cycles in which a timing error occurs.

For example, consider the circuit in Figure 2.2 consisting of two timing paths. P_1 toggles in cycles 2 and 4, and P_2 toggles in cycles 4 and 7. At voltage V_1, P_1 is at critical slack, and P_2 has 3 ns of timing slack. Scaling down the voltage to V_2 causes P_1 to have negative slack. Since P_1 toggles in 2 of 10 cycles, the error rate of the circuit is 20%. At V_3, the negative slack paths (now P_1 and P_2) toggle in 3 of 10 cycles, and the error rate is 30%. Note that although two paths violate timing in cycle 4, recovery for both violations happens simultaneously, so the cost of recovery is the same as it would be for a single violation.

2.1   EVAL and BlueShift

One common technique among several stochastic optimizations [18, 25, 26, 33] is to target the most frequently exercised paths in a stochastic design for special optimizations, since highly exercised paths have the potential to cause the most errors. EVAL [33] is a design-level optimization that aims to increase the amount that frequency can be overscaled in a timing speculative design. EVAL attempts to increase the efficiency of frequency overscaling by optimizing the most frequently-exercised paths in a design at the expense of the majority of the static paths. Consequently, errors are prevented on the highly active paths (that would cause many errors) and allowed on the infrequently exercised paths (that cause relatively few errors).

EVAL trades error rate for frequency by shifting, tilting, or reshaping the path slack distributions of the various functional units in a design.

BlueShift [18] is an application of EVAL that also optimizes a circuit for frequency overscaled operation. BlueShift uses EVAL techniques in an iterative optimization in an attempt to reduce the error rate of a circuit module. In each iteration, some optimizations are performed, and the error rate is checked. If the error rate is less than the target rate, the module optimization finishes. Otherwise, optimization iterations continue.

Each optimization step involves adding timing slack to the paths that cause the most timing errors and performing a gate-level simulation to check whether the path adjustments have brought the error rate below the target. BlueShift uses two methods to add slack to frequently-exercised circuit paths forward body biasing of selected gates and path constraint tuning that applies tighter timing constraints for selected paths.

BlueShift aims to improve performance by allowing the frequency of a circuit to be increased without incurring too many additional timing violations. BlueShift optimizes only the post-synthesis circuit over many layout iterations, Other stochastic optimizations target other

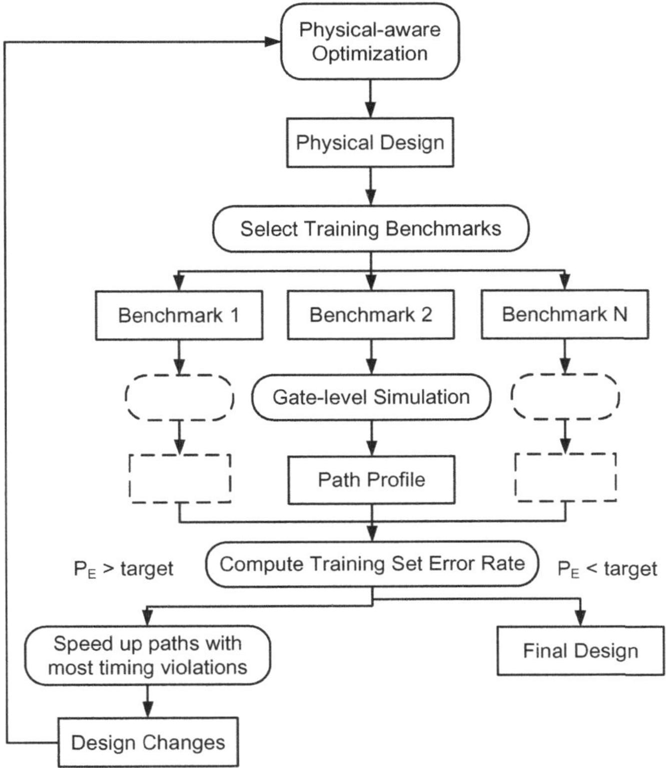

Fig. 2.3 BlueShift performs iterative optimization steps using the techniques of EVAL in order to reduce the error rate of a frequency overscaled circuit [18].

stages of the design flow such as synthesis and mapping. One limitation of BlueShift is the use of repetitive gate level simulations to obtain path profiles after making iterative design changes. This type of design cycle may easily become impractical, especially for large, modern SOC designs, where the number of post-sizing, layout, extraction simulation steps is often limited by runtime constraints. Figure 2.3 gives an overview of the iterative optimization performed by BlueShift.

2.2 Dynatune

Dynatune [43] is another stochastic optimization technique, similar to previously mentioned techniques, that improves performance by enhancing the efficiency of frequency overscaling-based timing speculation. Rather than treating all gates in a design as equals, Dynatune focuses optimization efforts on the most dynamically critical cells. The most dynamically critical cells are those with the highest switching activity. Dynatune starts with a circuit design that is implemented with high V_t cells (which have high delay and low leakage), and replaces some of the most dynamically critical cells with low V_t cells. This reduces their delay, but increases their leakage power consumption. Dynatune replaces as many cells as possible while staying below a specified threshold for leakage power. Since some paths that remain with many high V_t cells may not meet timing after optimization is finished, the design is timing speculative and must incorporate an error recovery mechanism to cover the case when timing speculation results in errors. Therefore, the goal of Dynatune is to assign low V_t cells, based on dynamic criticality, to maximize the peak frequency of a timing speculative design while staying within a given leakage power budget.

2.3   CRISTA

CRISTA [16] is a technique that targets an average case operating point by partitioning the paths of a design into two groups — paths with long (critical) delays and paths with short delays — then dynamically adapting the clock period of a circuit based on which paths are active. CRISTA uses Shannon expansion-based partitioning to isolate the critical paths in a circuit. Based on the dichotomy created by this partitioning, CRISTA downsizes cells on the most critical paths (thus increasing their delay and decreasing their power) and upsizes cells on the noncritical paths (thus affording them more timing slack at the expense of increased power). During operation, CRISTA uses additional logic to predict when paths in the long delay group will be exercised. Based on these predictions, CRISTA dynamically scales the clock frequency so that all paths are allowed enough time to execute in a given cycle — long paths are allowed multiple cycles to evaluate. Since paths in the short delay group are afforded extra timing slack during optimization, more overscaling can be done in the case when no long paths are excited. Thus, when activity on the long paths is infrequent, CRISTA can significantly improve performance in the average case.

To use CRISTA, a circuit must detect when a critical path will be excited, in order to allow extra cycles for path evaluation. Thus, CRISTA requires changes to the structure of the original circuit to isolate critical paths and detect which inputs will excite them. Since CRISTA requires circuit-specific optimization, alternative techniques that do not change the original circuit structure may be more easily applied to a wider range of circuits and excitation conditions.

On another note, CRISTA does not actually perform timing speculation. Rather than detecting and correcting timing errors, CRISTA predicts when a timing error would occur and scales circuit delay to avoid the error. In case CRISTA were used for timing speculation, the desired optimization would be different, since the longest paths in a design are not necessarily the ones that cause the most errors. It is only the longest paths with the highest activity that cause the most errors in a timing speculative design.

2.4 Better-Than-Worst-Case Synthesis

Like other stochastic optimizations discussed in this section, work on better-than-worst-case (BTWC) logic synthesis [11] has also proposed to use activity information to reduce the error rate of an overscaled design. Whereas traditional synthesis tools attempt to minimize delay for a logic block, a BTWC synthesis tool also consider switching probabilities when implementing a design. Reducing switching activity can result in fewer errors for an overscaled design.

BTWC logic synthesis uses functional information to reduce the probability of error for scenarios in which multiple possible logic decompositions for a block have equal cost. In such cases, ties between the original cost function (delay) are broken by a new cost function that takes switching probability into account. The tiebreaker cost function is a sum of delay, weighted by switching probability. Thus, the logically equivalent decomposition with the least switching activity (i.e., the most biased signal probabilities) is selected in the event of a tie.

Some limitations of BTWC synthesis in relation to other stochastic optimizations are the inability to target a specific error rate and potentially limited impact, since only logic with multiple candidate implementations that have the same delay and functionality are optimized.

2.5 Recovery-Driven Design

Conventional hardware is designed and optimized using techniques that aim to ensure correct operation of the hardware under all possible PVT variations. BTWC design techniques [1] save power by eliminating guardbands, but are still aimed at ensuring correct hardware operation under nominal conditions.

Recovery-driven design [26] contends that the use of error resilient design techniques should fundamentally change the way that hardware is designed and optimized. That is, given that mechanisms exist to tolerate hardware errors, rather than designing and optimizing hardware for correct operation hardware should be optimized for a target error rate, even during nominal operation. A recovery-driven design deliberately allows timing errors to occur during nominal operation, while relying on an error resilience mechanism to tolerate these errors.

The error rate target of a recovery-driven design is chosen such that any errors produced in the optimized design can be gainfully tolerated by hardware-based error resilience that detects and corrects errors in hardware [15, 17] or software-based error resilience in which errors are allowed to propagate from hardware and affect applications [20, 37, 39]. The expectation behind recovery-driven design is that the "underdesigned" hardware will have significantly lower power or higher performance than hardware optimized for correct operation. Also, because errors are allowed, the design methodology can exploit workload-specific information (e.g., activity of timing paths, architecture-level criticality of timing errors, etc.) to further maximize the power and performance benefits of underdesign.

Increasing the target error rate for a processor module increases the potential for power savings or performance gains, since the module can be operated at a lower voltage or a higher frequency. In practice, the target error rate of a recovery-driven design is chosen such that an error recovery mechanism can correct the errors that result from timing speculations and still reduce energy. Thus, the design methodology accounts for degradation in performance or output quality that may result from error recovery. Recovery-driven design manipulates the error distribution of a design through its slack distribution in order to increase the efficiency of computing in the face of timing variations. The energy benefits of exploiting error resilience are maximized by redistributing timing slack from paths that cause very few errors to frequently-exercised paths that have the potential to cause many errors. This reduces the error rate at a given voltage or frequency, and hence reduces minimum supply voltage and power or increases maximum frequency and performance for a target error rate.

The goal of recovery-driven design in the context of voltage overscaling can be stated formally as follows. Given an initial netlist N_0, a set of cell libraries characterized for allowable operating voltages, toggle rates for the toggled paths in the netlist, and a target error rate ER_{target}, produce the optimized netlist $N_{V_{\text{opt}}}$ and operating voltage V_{opt} that minimize the total power consumption $W_{V_{\text{opt}}}$ of the circuit, such that the error rate of the optimized netlist does not exceed ER_{target}. Figure 2.4 demonstrates the goal.

Recovery-driven design work [26] proposes a heuristic implementation based on slack redistribution. The heuristic targets energy reduction with a two-pronged approach — extended voltage scaling through cell upsizing on critical and frequently-exercised circuit paths (*OptimizePaths*), and leakage power reduction achieved by downsizing cells in noncritical and infrequently-exercised paths (*ReducePower*). The heuristic searches for the combination of these two techniques that provides the lowest total power consumption for a circuit, by performing path optimization and power reduction at each voltage step and then choosing the operating power at which minimum power is observed.

Figure 2.5 illustrates the evolution of a slack distribution throughout the stages of the power minimization procedure. Each iteration begins as voltage is scaled down by one step (a). After partitioning the paths into sets containing positive and negative slack paths, *OptimizePaths*

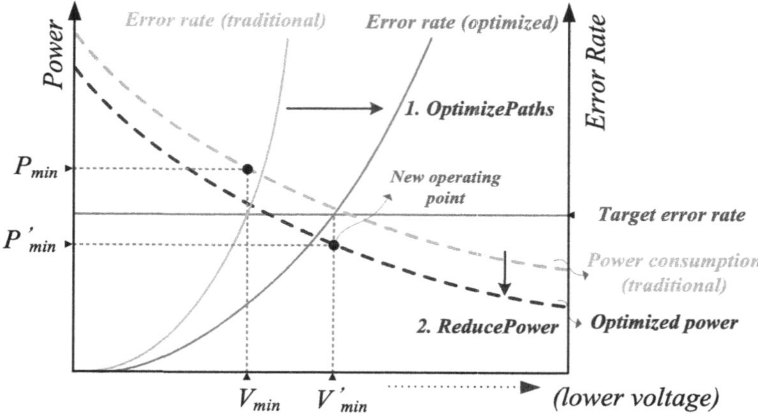

Fig. 2.4 A recovery-driven design flow redistributes slack from infrequently-exercised paths to frequently-exercised paths and performs cell downsizing to accomodate average-case conditions. These optimizations reduce the power consumption of a circuit and extend the range that voltage can be scaled before a target error rate is exceeded. The combination of these factors produces a design with significantly reduced power consumption [26].

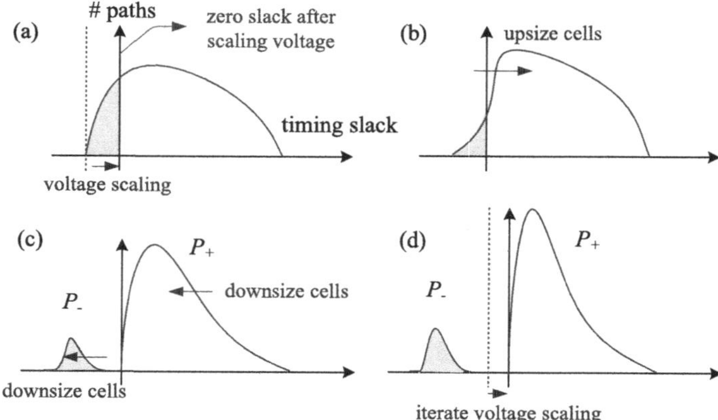

Fig. 2.5 The power minimization heuristic for recovery-driven design reshapes a circuit's path slack distribution by redistributing slack from paths that rarely toggle to paths that toggle frequently. This extends the amount of ovescaling that can be done for a given error rate target [26].

attempts to reduce the error rate by increasing timing slack on negative slack paths (b). Next, the heuristic allocates the error rate budget by selecting paths to be added to the set of negative slack paths, and downsizes cells to achieve area/power reduction while respecting the dichotomy between negative and nonnegative slack paths (c). This cycle is repeated over the range of potential operating voltages to find the minimum power netlist and corresponding voltage (d). In Figure 2.5, P_+ is the set of paths that must have nonnegative slack after power reduction, and P_- is the set of paths that are allowed to have negative slack.

Figure 2.6 presents the algorithmic flow of a recovery-driven design heuristic that targets power minimization [26]. The heuristic

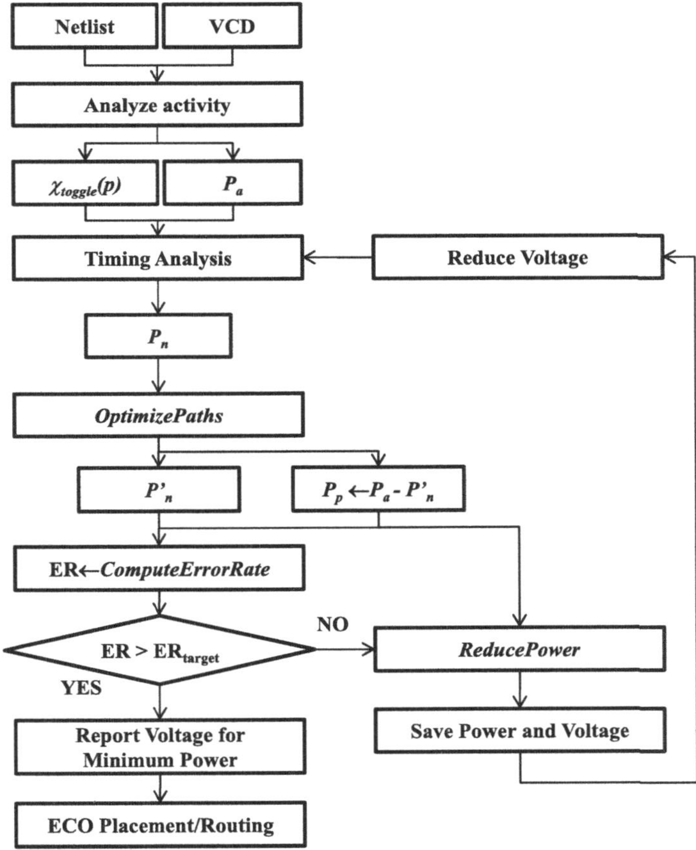

Fig. 2.6 Algorithmic flow of a heuristic for minimizing power for a target error rate. P_a is the set of all paths toggled during simulation. P_p is the set of all nonnegative slack paths. P_n is the set of all negative slack paths in P_a. $\chi_{\text{toggle}}(p)$ is the set of cycles in which path p is toggled [26].

couples path optimization to extend the range of voltage scaling (*OptimizePaths*) with area minimization to achieve power reduction (*ReducePower*). The details of the heuristic are in Figure 2.6 and [26].

By optimizing a design to achieve maximum efficiency at a nonzero error rate rather than for correct operation, recovery-driven design has demonstrated significantly increased energy efficiency for voltage overscaling-based timing speculation [26]. Figure 2.7 compares recovery-driven design for a Razor-based [15] timing speculative processor against several alternative design styles. Conventional P&R and Tight P&R are highly optimized conventional CAD flows with different timing constraints. PowerOpt X% denotes a recovery-driven design that targets an error rate of X%, and SlackOpt denotes a gradual slack design [25]. PCT denotes a path constraint tuning approach similar to the technique used by BlueShift [18]. Figure 2.7 demonstrates that adopting a recovery-driven design style enables additional energy savings over competing techniques, due to extended overscaling and lower area and leakage. Designing the processor for the error rate target at which Razor operates most efficiently allows the recovery-driven design to extend the range of voltage scaling from 0.84V for the best "designed for correct operation" processor to

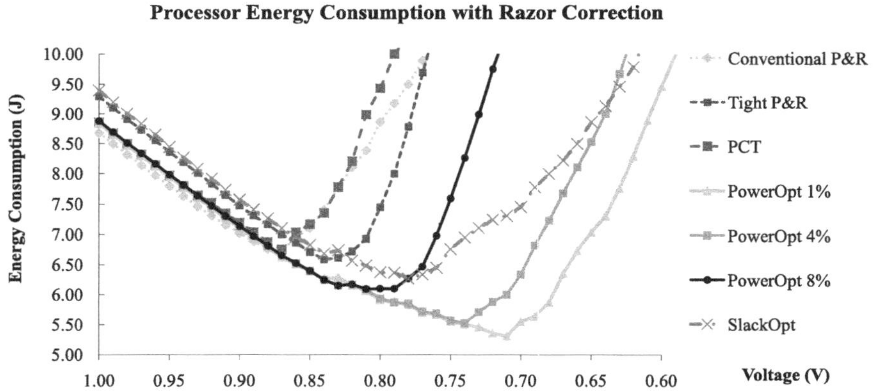

Fig. 2.7 By optimizing the processor for a certain error rate, then correcting the errors with an error-resilience mechanism, recovery-driven design achieves additional energy savings over competing approaches. Results are shown for a processor that uses Razor-based timing speculation. The additional energy benefits of the recovery-driven design are due to extended overscaling and reduced area and leakage.

0.71V for the recovery-driven design optimized for an error rate of 1%, affording an additional 19% energy reduction.

2.6 Gradual Slack Design

Gradual slack design [27] is an extension of recovery-driven design that reshapes the slack distribution of a design to create a gradual failure characteristic, rather than the typical critical wall. While error rate-optimized, recovery-driven designs achieve better energy efficiency at a single target error rate, gradual slack designs have the ability to trade reliability, throughput, or output quality for energy savings over a range of error rates. Gradual slack design techniques are used to create *soft processors* [25] – processors that degrade gracefully under variability. Figure 2.8 describes the optimization approach for gradual slack design.

The recovery-driven design flow presented in Section 2.5 selects paths that are allowed to make errors in order to optimize for a specific target error rate. To achieve a graceful failure characteristic, a gradual slack design flow instead optimizes for a maximum target error rate corresponding to the desired range of scalability over which the design should make efficient performance/power tradeoffs. For gradual

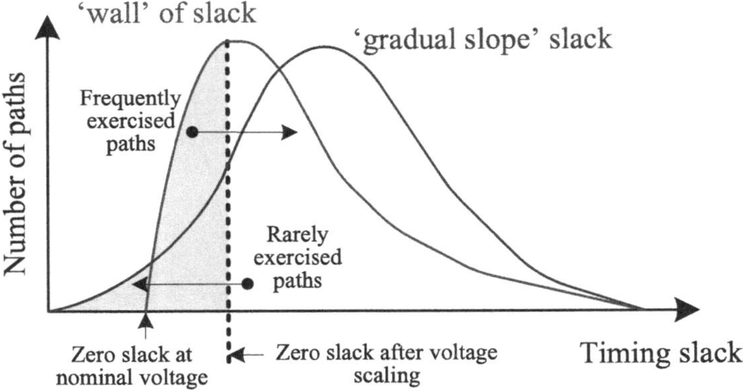

Fig. 2.8 The goal of a gradual slack [27] or soft processor [25] design is to transform a slack distribution characterized by a critical "wall" into one with a more gradual failure characteristic. This allows performance/power tradeoffs over a range of error rates, whereas conventional designs are optimized for correct operation and recovery-driven designs, are optimized for a specific target error rate.

slack design, only the negative slack paths in the scaling range with the highest switching activity are optimized, to maximize the range of voltage scalability for the target range of error rates. Cells that have negligible switching activity are downsized, such that the slack distribution for the active paths and the error rate of the processor are not affected. In this way, a slack distribution with a critical wall is transformed into one with gradual sloping slack. A gradual slack design would not typically have a cluster of paths in the negative region of the slack distribution (like a recovery-driven design might), unless the target range of scalability only includes nonzero error rates.

By optimizing for a range of error rates, gradual slack design ensures that there are not too many highly exercised critical paths that cause the error rate to increase rapidly. Such behavior would prevent additional energy savings when a higher error rate is allowable, and would thus prevent efficient energy/performance tradeoffs. Figure 2.9 compares the processor (OpenSPARC T1) error rate for several competing design styles described in Section 2.5. By increasing slack on several highly exercised paths, gradual slack design achieves a lower error rate for a given voltage, which translates into a larger range of voltage scaling for a given error rate target. Note that tightly constrained P&R

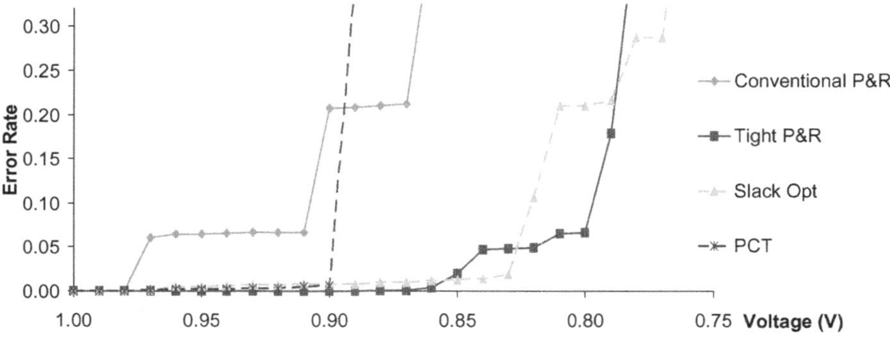

Fig. 2.9 A gradual slack design [25, 27] achieves a lower error rate for a given voltage than most competing design styles. This allows more overscaling for a given target error rate. Although a conventional CAD flow that tightly constrains all paths and thus highly optimizes them may achieve similar error rate behavior, a gradual slack design, which is functionally aware, optimizes a smaller set of critical paths and thus has considerably lower area and power than a highly optimized design.

also achieves an extended range of voltage scaling. However, since the conventional design optimization approach is not functionally aware, the power consumption of Tight P&R is much higher than that of SlackOpt. Consequently, SlackOpt is more efficient for trading performance and energy over a target range of error rates.

3

Architecture Frameworks for Stochastic Computing

At the system level, several stochastic architecture frameworks have been proposed. These architecture frameworks are structured around lower-level stochastic computing techniques, which they exploit to increase system-level efficiency in the face of low-level nondeterminism. Often, to achieve efficiency, these frameworks must adapt to varying reliability requirements for different workloads, environmental conditions, and utilization patterns. This section discusses architectural frameworks that dynamically build upon stochastic computation and optimization techniques to achieve system-level energy efficiency in spite of inherent low-level nondeterminism.

3.1 Error-resilient System Architecture (ERSA)

ERSA [29] (Figure 3.1) is an asymmetric multi-core architecture that contains cores with varying degrees of reliability. Each core belongs to one of two classes — Super or Strict-reliability cores (SRC) or Relaxed-reliability cores (RRC). Typically, a single SRC is responsible for maintaining acceptable behavior for the processor. The SRC executes control-intensive code that is not tolerant to errors, schedules

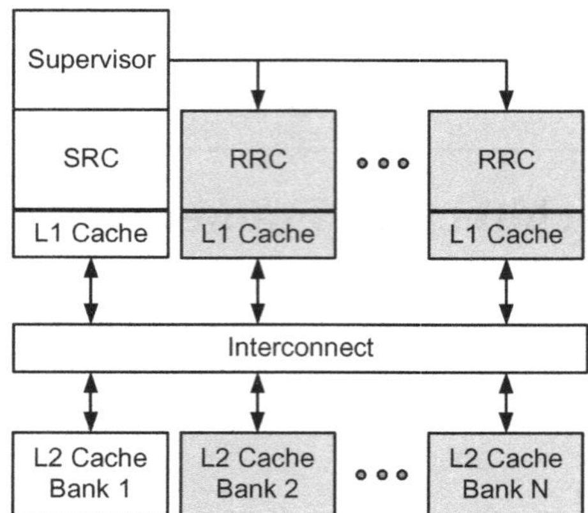

Fig. 3.1 ERSA [29] is an asymmetric multi-core that contains both reliable (SRC) and reduced-reliability (RRC) cores. Error-tolerant threads run on RRCs, while computations that cannot tolerate errors are performed on the SRC.

tasks on the RRCs, and checks for errors resulting from illegal memory accesses and timeouts of the RRCs.

The RRCs, are designed for reduced reliability targets, perhaps employing some of the techniques discussed in the previous section. These cores are the main source of throughput for an ERSA multi-core. Because of their relaxed correctness constraints, RRCs may be significantly less demanding of system or design resources than their counterpart SRCs. To enhance the feasibility of computing primarily on RRCs, software designed for ERSA should be built around error-resilient algorithms.

3.2 Algorithmic Noise Tolerance (ANT)

Algorithmic noise tolerance [20] proposes the use of low-energy *soft DSP*. A soft DSP design uses voltage overscaling or better than worst case design to reduce energy consumption at the expense of some timing errors. Faced with these errors, ANT uses knowledge of the expected output signal distributions to qualify results and filter out outliers. Specifically, ANT employs a reduced complexity estimator block in parallel with the main circuit block (Figure 3.2). The estimator block computes a reduced precision, but error-free, version of the output,

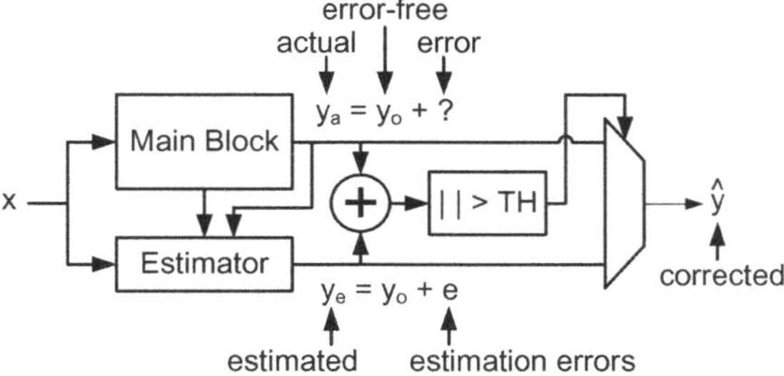

Fig. 3.2 In the ANT [20] framework, a reduced precision estimator block performs computations in parallel with the main circuit block. Comparing the outputs of the two blocks indicates when an error has occurred in the main block. In this case ANT uses the estimated output to reduce the impact of errors on output SNR.

which is compared against the output of the main block. A large difference between the outputs indicates that a timing error has occurred in the main block. In the case that an outlier is detected, ANT discards the erroneous main block output and uses instead the reduced precision version of the output. By reducing voltage and mitigating the effect of erroneous computations on output quality, ANT enables reduced energy consumption while minimizing degradation in output SNR.

3.3   Stochastic Sensor Network on Chip

A stochastic sensor network-on-a-chip [42] (SSNOC) is a communication-inspired design style that aims to provide variation tolerance in a noisy environment. Using robust estimation techniques a SSNOC tolerates errors induced by process variation and voltage overscaling. Due to localized sources of variation, such as particle strikes, thermal hotspots, and process variations, a single, centralized computation resource may be vulnerable to static and dynamic nonidealities. To overcome this vulnerability, a SSNOC architecture decomposes a centralized computation resource into a network of statistically similar [42] sensors. Statistical similarity implies that although individual sensor reading may contain errors, the average value of each sensor equals the expected value of original computation resource. To reduce implementation cost, the distributed sensors are

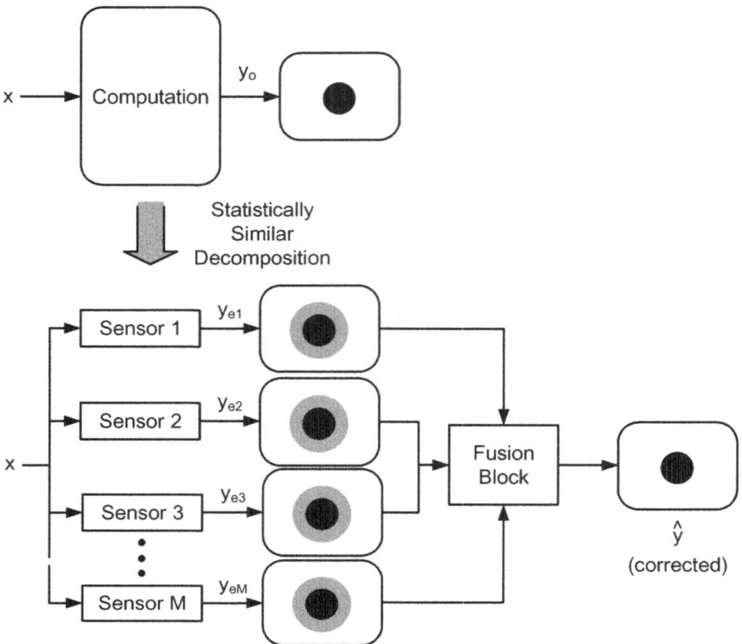

Fig. 3.3 The SSNOC [42] architecture decomposes a computation resource into a network of statistically similar sensors. Though individual sensors readings may be erroneous, the fusion block uses robust estimation techniques to combine the sensor values in order to produce a variation-tolerant output that is close to the expected output.

designed to have reduced complexity. The outputs of the sensors are fused together by a fusion circuit block to produce the final output, using principles of robust estimation theory (Figure 3.3).

Two sources of errors affect SSNOC computations — estimation errors due to the reduced precision of the distributed sensors and errors induced by process and dynamic variations. Although the mean value of the estimation error is expected to be zero, the distribution of variation-induced errors is unknown. Therefore, there is a random variation component in the final output of the SSNOC.

The fusion block in a SSNOC architecture is responsible for combining sensor outputs to produce the final output of the network, using robust estimation. Depending on the type of computation being performed, the fusion block may take on a different form. This brings to light some potential disadvantages of SSNOC. Namely, the computation must be decomposable into a distributed network of statistically similar sensors, and the fusion block used to combine sensor data must be custom designed for each SSNOC. In cases where a SSNOC implementation is possible, error detection probability can be significantly improved, due to hardware replication and distribution.

3.4 Variation-Adaptive Stochastic Computer Organization (VASCO)

VASCO [38] is a soft processor architecture that targets energy-efficient computation in the face of hardware variations and dynamically changing compute conditions. VASCO employs a range of different design styles and stochastic optimizations, including recovery-driven design [26], gradual slack design [25], and timing speculative memory. For a mixture of robust and error-sensitive applications, VASCO

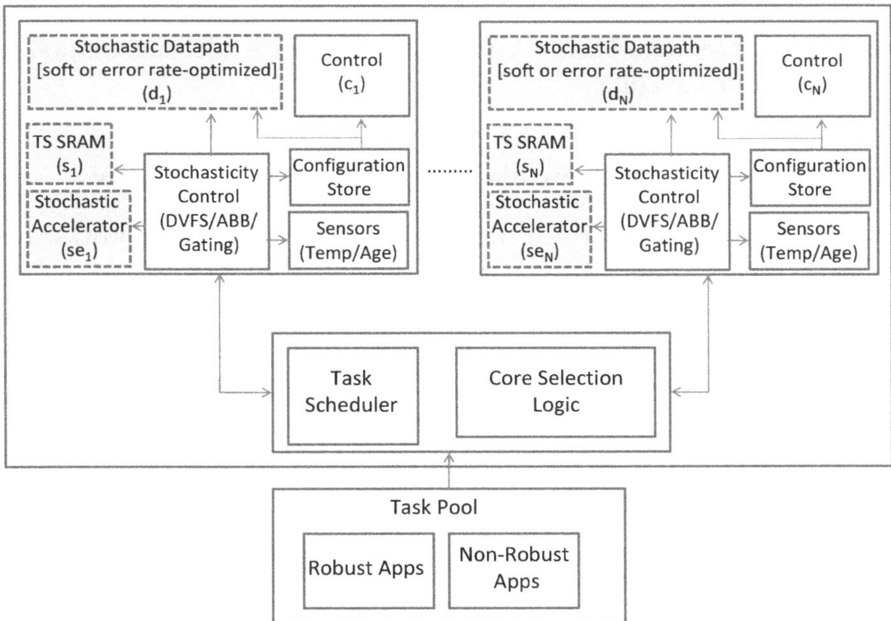

Fig. 3.4 The VASCO architecture [38] adapts its hardware reliability based on workload characteristics and environmental conditions to maximize the energy and performance benefits of exploiting stochasticity.

achieves energy efficiency by dynamically adapting the processor configuration and hardware reliability to match compute conditions.

VASCO is a tiled multi-core architecture. At the heart of each compute tile is a stochasticity control unit that adjusts the reliability of the tile and decides how to execute tasks based on the state of the hardware and the requirements of the tasks mapped to the tile. Tasks may be executed either on a stochastic datapath or a stochastic accelerator, depending on their characteristics. Although control logic on the tiles is strictly reliable, the reliability of the stochastic datapath can be adjusted according to the reliability requirements of a task. For example, the stochasticity control unit administers such parameters as voltage, frequency, body biasing, and gating, and can adjust the error rate of the stochastic datapath and other stochastic on-chip components to trade reliability for energy savings. Sensor data, such as temperature and aging, is also fed to the stochasticity controller and used in deciding how to adapt controlled parameters for maximum energy efficiency. The more variability or errors a task is able to tolerate, the more energy VASCO can save by exploiting adaptable stochastic optimizations.

3.5 Architectural Principles for Stochastic Processors

Previous sections described different architectural frameworks for stochastic computing. Recent work [35] attempts to uncover the fundamental reasons why conventional architectures may be inefficient for stochastic computing, and uses the insights gained to propose optimizations for stochastic architectures. The work identifies three main reasons why simply adding an error resilience mechanism and allowing errors in a conventional architecture may be inefficient. While design-level techniques focus on manipulating error rate through the slack distribution, architecture-level adaptations are able to influence both the slack and activity distributions of a design. Thus, they may have increased effectiveness over design-level techniques.

Architecture-level stochastic optimization work [35] demonstrates that the error distribution of a design depends strongly on architecture and that the error distribution of a design that has been architected for error-free operation may limit scalability and energy efficiency for BTWC operation. Thus, optimizing architecture for correctness can result in significant inefficiency when the actual intent is to perform timing speculation. On the other hand, architectural optimizations can be used to increase the efficiency of a design that employs timing error resilience. In other words, one would make different, sometimes counter-intuitive, architectural design choices to optimize the error distribution of a design to exploit timing error resilience.

Figure 3.5 demonstrates that microarchitecture can have a significant effect on error rate. The figure shows the error rate behavior for four different microarchitectural configurations of the FabScalar [10] processor. Each microarchitecture has a significantly different error rate behavior, demonstrating that slack, activity, and error rate indeed depend on microarchitecture. Differences in the error rate behavior of different microarchitectures are due to several factors. First, changing the sizes of microarchitectural units like queues and register files changes logic depth and path delay regularity, which in turn effects the slack of many timing paths. Second, varying architectural parameters such as superscalar width has a significant effect on logic complexity [30]. Complexity, fanout, and capacitance change path delay sensitivity to voltage scaling and cause the shape of the slack distribution to change. Finally, changing the architecture alters the activity distribution of the processor, since some units are stressed more heavily,

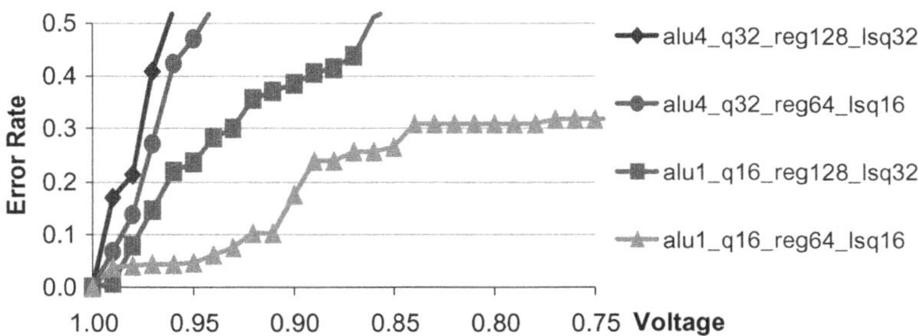

Fig. 3.5 Different microarchitectures exhibit different error rate behaviors, demonstrating the potential to influence the energy efficiency of a timing speculative architecture through microarchitectural optimizations. (Notations describe the number of ALUs, issue queue length, number of physical registers, and load store queue size for a microarchitecture.)

depending on how the pipeline is balanced. High activity in units with many critical paths can cause error rate to increase more steeply. Likewise, an activity pattern that frequently exercises longer paths in the architecture limits overscaling.

As noted above, certain architectural features may degrade the energy efficiency of a timing error resilient processor. Thus, when optimizing the microarchitecture of a timing error resilient processor, certain design principles should be observed.

Typically, energy-efficient processors devote a large fraction of die area to structures with very regular slack distributions, such as caches and register files. These structures typically have high returns in terms of energy efficiency (performance/watt) during correct operation. For example, 75%–80% of the critical paths in the Alpha EV7 reside in the L1 caches and register files (Figure 3.6) [31].

While regular structures are architecturally attractive in terms of processor efficiency for correct operation, such structures have slack distributions that allow little room for overscaling. This is because all paths in a regular structure have similar lengths, and when one path has negative slack, many other paths also have negative slack. For example, consider a cache. Any cache access includes the delay of accessing a cache line, all of which have nearly the same delay. So, no matter which cache line is accessed, the delay of the access path will be nearly the same. Compare this to performing an ALU operation, where the delay can depend on several factors including the input operands and the operation being performed. When exploiting timing error resilience for energy reduction, the energy-optimal error rate is found by balancing the marginal benefit of reducing the voltage with the marginal cost of

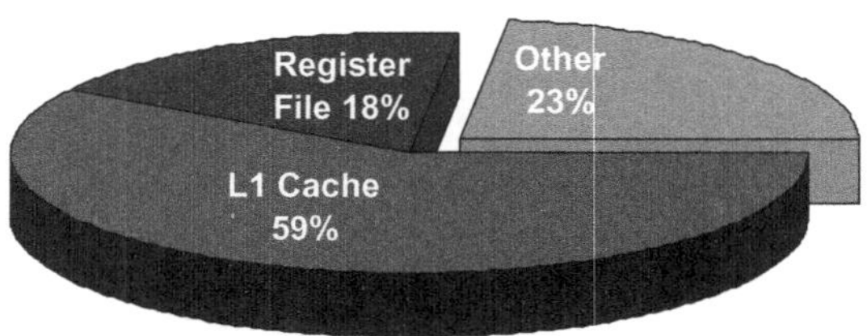

Fig. 3.6 Typically, slack distributions of processors are dominated by regular structures. Caches and register files account for a large fraction of the critical paths of a processor [31].

recovering from errors [15]. When many paths fail together, error rate
and recovery overhead increase steeply upon overscaling, limiting the
benefits of timing speculation. Reducing the number or delay of paths
in a regular structure can reshape the slack distribution, enabling more
overscaling and better timing speculation efficiency.

For an example Alpha core [41], the register file is the most regular
critical structure. Figure 3.7 shows slack distributions for the Alpha
core with different register file sizes. As the size of the register file
increases, the regularity of the slack distribution also increases, as does
the average path delay. Figure 3.7 confirms that the spread of the slack
distribution decreases with a larger register file. Additionally, path slack
values shift toward zero (critical) slack due to the many critical paths
in the register file. Table 3.1 shows standard deviation and mean values
for the slack distributions of the processors with different register file
sizes. The table confirms that regularity (represented by the standard
deviation of slack) increases, and average slack decreases with the size
of the register file. (Note that smaller σ_{slack} means a more regular slack
distribution.) Changing the cache size affects the slack distribution in
a similar manner. For example, σ_{slack} reduced by 25% for the Alpha

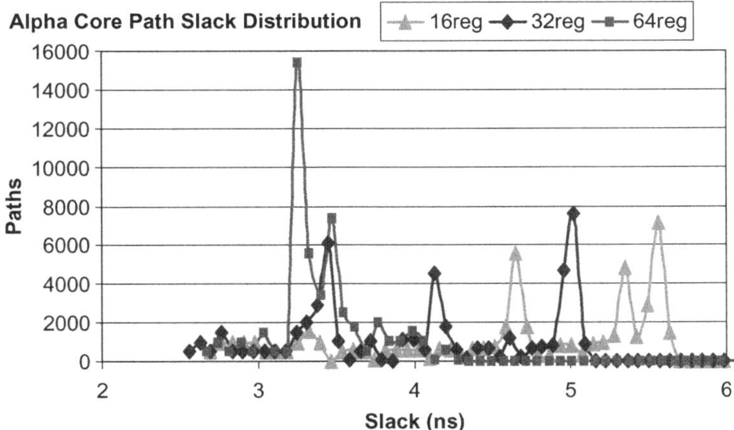

Fig. 3.7 The sizes of regular structures can significantly influence the slack distribution of
a microarchitecture. Reducing the size of the register file (a regular structure) increases
the spread of the slack distribution, resulting in fewer paths bunched around the point of
critical slack.

Table 3.1.   Mean and standard deviation of path slack, relative to the clock period, for an Alpha processor with different register file sizes.

	16 reg	32 reg	64 reg
μ_{slack}	46%	41%	34%
σ_{slack}	10%	9%	6%

core and 23% for an example MIPS core [41] when the cache size was increased from 2 KB to 4 KB.

Architectural design decisions that reshape the slack distribution by devoting less area to regular structures or moving regular structures off the critical path can enable more overscaling and increase energy efficiency for timing speculative processors. In other words, additional overscaling enabled by architectures with smaller regular structures can outweigh the energy benefits of regularity when designing a resilience-aware architecture. Since regularity-based decisions may also impact power density, yield, and performance, architectural decisions should consider these constraints in addition to the chosen optimization metric.

Typically, processors are architected for energy efficiency during error-free operation at a single power/performance point and are not expected to scale to other points. However, timing speculative architectures achieve benefits by scaling beyond the normal, worst case operating point to eliminate conservative design margins. The change in the shape of the slack distribution as voltage changes depends on the delay scalability of the paths — i.e., the rate at which path delays change as voltage is scaled. Therefore, unlike conventional architectures, architectures optimized for timing speculation should consider the delay scalability of different microarchitectural structures.

There are several architectural characteristics that affect delay scalability that conventional processors ignore to varying degrees. One factor that affects delay sensitivity to voltage scaling is logic complexity. In a conventional processor, microarchitectural components are optimized largely oblivious to complexity, as long as the optimization improves processor efficiency at the nominal design point. However, more complex structures with more internal connections, higher fanouts, deeper logic

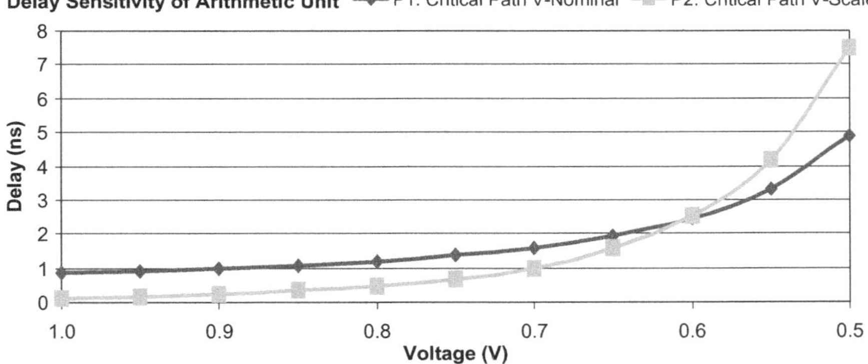

Fig. 3.8 Some paths are more sensitive to voltage scaling than others. Complex logic with many high fanout paths like P2 can limit overscaling in a timing speculative architecture, since delay increases more quickly for such logic as voltage is scaled down.

depth, and larger capacitance are more sensitive to voltage scaling, potentially limiting overscaling for a timing speculative processor.

Figure 3.8 demonstrates how path delays scale at different rates as voltage is scaled down on the ALU of the OpenSPARC T1 [40] processor. Path P1 is the critical path at nominal voltage. However, the delay of P2 is more sensitive to voltage scaling due to increased fanout. At a particular voltage, the critical path of the design actually changes due to the imbalanced delay sensitivities. The slack distribution of a processor with many complex logic structures becomes more critical more quickly as voltage is scaled, limiting overscaling.

To maximize the energy efficiency benefits of timing speculation, architectural decisions should be scalability-aware. For example, complex architectural structures with high degree of fanout should be optimized to reduce complexity, if possible. Similarly, less complex implementations of architectural units can be chosen when performance is not significantly impacted. Example optimizations include changing superscalar width and queue sizes — factors that strongly influence logic complexity. The capacitance of a logic structure also influences the rate at which delay increases with voltage reduction. If the impact on processor efficiency is acceptable, less area should be devoted to

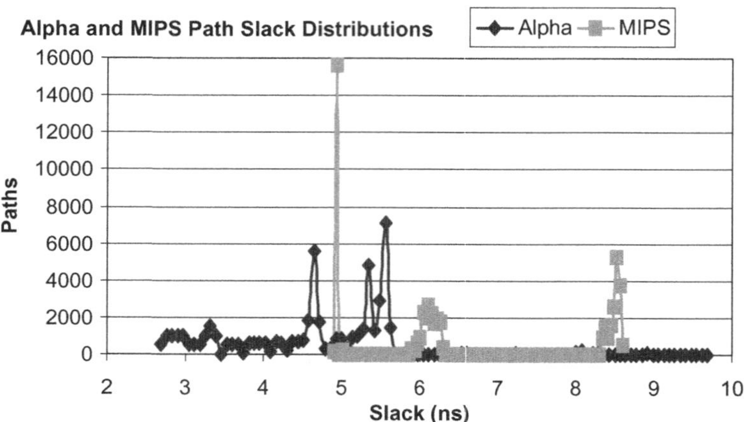

Fig. 3.9 The reduced regularity and complexity of the MIPS architecture, compared to the Alpha architecture, results in a slack distribution with greater average slack and reduced regularity.

complex, centralized structures with high internal capacitance (e.g., rename logic, wakeup/select logic, bypass logic, etc.).

Comparing the Alpha and MIPS architectures again reveals how architectural changes affect the slack distribution. Figure 3.9 compares the slack distributions of the MIPS and Alpha processors. The MIPS slack distribution has both higher mean and standard deviation than the distribution for the Alpha processor (61% higher mean and 57% higher standard deviation), indicating reduced regularity and complexity. These factors can be attributed to reduced word length, simpler ALU design, smaller area devoted to the register file, and a simpler, smaller instruction set, which results in less complex control logic throughout the processor.

Modern processors consistently employ architectural techniques such as pipelining, superscalar processing, caching, etc. to improve utilization by reducing the number of control and data hazards and mitigating long latency memory delays. In general, when designing for correctness, architectural design choices that increase utilization are desirable, as higher utilization of a processor core often leads to better performance. However, architectures with highly utilized critical paths are susceptible to high error rates, since increased activity on negative slack paths means more frequent errors. Architectural optimizations that reduce the activity of critical paths have the potential to reduce the error rate when timing speculation is performed.

As noted above, architectural parameters that have a strong impact on logic complexity, such as superscalar width [30], should be carefully optimized in an error-resilient processor to allow efficient scalability to multiple operating points. In addition, changing the superscalar width can significantly impact the activity distribution of a processor. Changing the MIPS architecture from a scalar design to a superscalar (*width* = 2) design increases average activity by up to 25%.

Note that activity reduction has associated costs in terms of throughput during correct operation. Thus, it is unwise to reduce activity at all costs. Rather, balancing the error rate reduction and energy efficiency benefits of activity reduction with the throughput benefits of high utilization produces the most efficient design.

4

Compiler Optimizations for Stochastic Computing

Sections 2 and 3 discuss how design-level optimizations and architecture-level frameworks can enable enrgy-reliability tradeoffs and improve the energy efficiency of timing error-resilient processors. In this section, we discuss how optimizing the software that runs on timing error-resilient processors can also improve their energy efficiency by manipulating their activity distributions [36].

As described in Section 2, the magnitude of energy efficiency benefits available from exploiting timing speculation (TS) depends on two factors — (a) *where* and (b) *how often* the processor produces errors when operating at an overscaled voltage or frequency. The path slack distribution of a timing speculative processor determines which paths cause errors when they are toggled. Likewise, the activity distribution of the processor describes how often paths are toggled, and thus determines the frequency of errors caused by a path when it has negative slack. Together, the slack and activity distributions dictate the error distribution of a processor, i.e., the locations and frequencies of errors produced in an overscaled processor.

Altering the error distribution of a timing speculative processor has the potential to increase the energy benefits of exploiting error resilience. As discussed above, architecture and design-level works have demonstrated that modifying the slack and activity distributions of a processor can increase the energy efficiency [25, 26, 27, 35] and performance [18, 33] of a timing speculative design. Since the program binary, in conjunction with the processor architecture, determines a processor's activity distribution, *timing speculation-aware compilation* [36] can also significantly increase the energy efficiency of a design that exploits timing error resilience.

The goal of TS-aware binary optimizations is to minimize the energy consumption of a timing speculative processor by manipulating its activity distribution to reduce the error rate for a given voltage and frequency or reduce the cost of error recovery. Since altering the activity distribution of the processor may also affect throughput, both the performance and power implications of TS-aware compilation should be considered during binary optimization. TS-aware binary optimizations fall into one of the following four categories.

4.1 Critical Path Avoidance

The first category of TS-aware optimization techniques change the set of frequently exercised paths in a processor to avoid activating the longest paths. Since these paths are the first to have negative slack when the processor is overscaled, throttling their activity reduces early onset timing violations.

In any processor, the set of paths that are exercised depends on the mix of instructions and data. For several hardware structures, the most timing critical paths may only be excited when certain instruction or data patterns are presented. Consider, for example, a ripple carry adder (RCA). The critical path of the RCA is excited only when a carry chain propagates from the least significant bit to the most significant bit. On average, the length of the dynamically critical path (the path that determines propagation delay in a given cycle) may be significantly shorter than the static critical path [34]. Optimizing the instruction stream can *prevent* activity patterns that exercise the static critical paths of hardware structures, resulting in more timing slack, on average, and more room for overscaling.

4.2 Activity Throttling

The second category of techniques reduce error rate by throttling activity in structures of the processor that cause the most errors. A program binary, in conjunction with the processor microarchitecture, determines how often different processor units are stressed. When a structure that exercises its critical paths for most instructions also has high utilization, the structure generates many errors when the processor is overscaled. Thus throttling the activity of such structures can significantly reduce the processor error rate for a given voltage, allowing the processor to scale to a lower voltage for a given error rate. It should be noted that activity throttling may lead to throughput degradation. However, if the reduction in power due to additional overscaling outweighs the loss in throughput, activity throttling increases overall energy efficiency.

4.3 Overlapping Errors

The previous categories deal with error avoidance. Another way to increase TS efficiency is to reduce the cost of error recovery. For many error recovery mechanisms, such as rollback, flushing, counterflow, and replay, when multiple timing errors occur in the same cycle, a single error recovery process can correct all the errors. The third category of compiler optimizations for stochastic computing focuses on overlapping the occurrence of timing violations so that multiple errors share a single error recovery process.

Since certain instruction and data combinations are more error prone than others because of how they interact with hardware, the goal of these optimizations is to group instructions based on their operands and their likelihood of causing a timing violation. If a program makes k errors while executing on a TS processor, the goal is to minimize the cost of recovering from the k errors so that the energy required to execute the program is minimized.

4.4 Recovery Cost-Aware Scheduling

The final category of techniques aim to schedule accesses to processor structures such that activity is inversely proportional to error recovery cost. This strategy aims to utilize timing speculation more efficiently by avoiding high cost errors. For example, with an error recovery mechanisms such as pipeline flushing or counterflow pipelining [15], recovery cost increases with distance from the front of the pipeline. This is because more pipeline stages must be flushed when a timing violation occurs. Optimizing the binary to reduce activity in the processor backend or shift activity from later stages to earlier stages may increase TS efficiency for such recovery mechanisms.

4.5 Optimizing the Binary to Reduce Error Rate

The exact optimization techniques that are most effective for a processor depend on the processor architecture and the exact modules of the processor that cause the most errors. Different pipeline stages cause errors at different rates, depending on their slack and activity distributions. Figure 4.1 shows the static slack distributions for the pipeline stages that cause the most errors in one specific superscalar pipeline implementation (FabScalar [10]). Figure 4.1 demonstrates that two stages in particular — the issue queue (IQ) and the load store unit (LSU) have the highest number of critical paths. Based on Figure 4.1, one might expect that IQ, having many more critical paths than all

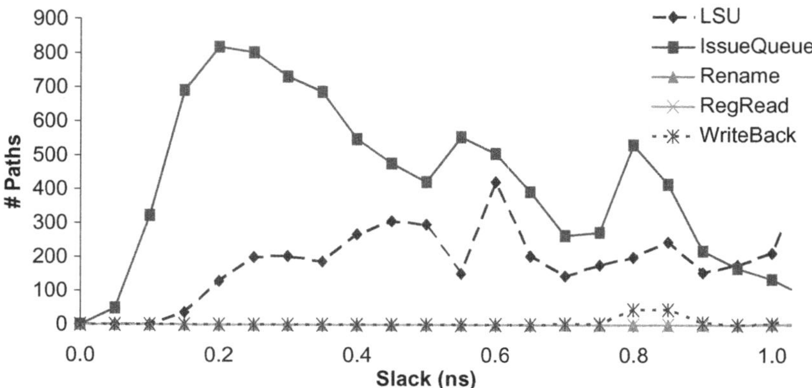

Fig. 4.1 The static slack distributions for the pipeline stages show how many critical paths they have. The static slack distribution does not, however, provide information about how often the paths toggle, which is essential in characterizing error rate.

other modules, would produce the most errors in the processor. However, the static slack distribution only shows the *potential* for paths to cause errors. To know how many errors the paths actually generate, their activity must also be known. Figure 4.2 shows activity-weighted, *dynamic* slack distributions that quantify the sum of toggle rates for all the paths at each value of timing slack (activity from SPEC benchmarks). The more timing critical *activity* a module has, the more errors it is likely to cause. From Figure 4.2, it is clear that the LSU dominates the error distribution of the FabScalar architecture.

We now describe the implementation of the LSU in the FabScalar processor to understand the dependence of error rate on program characteristics. The LSU (Figure 4.3) performs memory disambiguation for

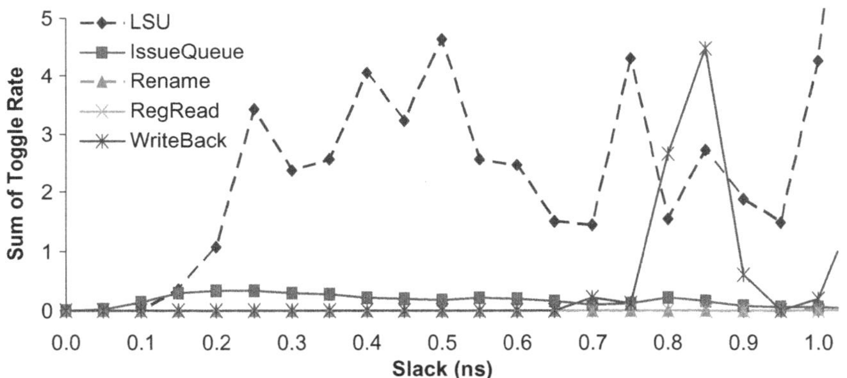

Fig. 4.2 The activity weighted (dynamic) slack distributions for different pipeline stages indicate how much timing critical activity they have, and by extension, how frequently they will produce errors for a given level of overscaling.

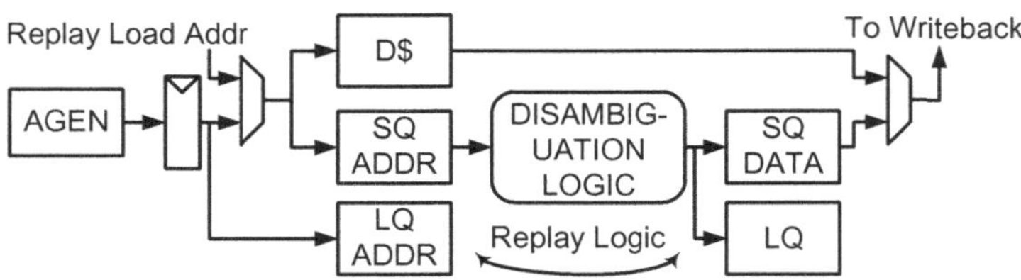

Fig. 4.3 Memory disambiguation is on the critical path of the LSU [10]. The path delay is longest when store-to-load forwarding is required, since this necessitates an access to the SQ data RAM, in addition to the other disambiguation operations.

the processor. This involves checking for dependencies between loads and stores. After address resolution, a store must search the address CAM of the LQ and process all entries with matching addresses to determine if any load issued out of order and broke a RAW dependence. Load disambiguation is more complicated because it includes store-to-load forwarding. In addition to the search through the SQ address CAM, a load must generate a mask vector indicating all preceding stores in program order. Matching entries from the CAM search are filtered by the mask vector, and the latest resulting entry, if any, forwards data from the SQ RAM to the load.

LSU delay depends on program characteristics for several reasons. The primary reason is that the store-to-load forwarding path is on the static critical path of the LSU. Since many RAW dependencies in the code lead to more forwarding, the timing error rate will be higher for code with a relatively large number of RAW dependencies. Program characteristics also determine the utilization of the LQ and the SQ, which, in turn, dictates access delays. For example, when the LQ or SQ are nearly full, as may be common for memory-centric codes, more entries must be accessed in a single cycle to generate mask vectors. This increases the length of the propagation path, and consequently, increases delay. Additionally, when there are many dependencies between memory operations, address CAM searches generate many hits, increasing load capacitance and delay for the CAM access. Finally, propagation delay increases when many hits are signaled in parallel (due to many potential dependencies), since the average length of the propagation path from the CAM entries to the port increases. Hence, the average delay is higher for memory-centric code as well as for code with a large number of dependencies.

As described above, activity on the static critical paths of the LSU can be reduced by avoiding dependent memory operations and scenarios that cause the LSQ to fill up. This can enable significantly deeper voltage overscaling, since the LSU is the source of many timing violations.

Loop unrolling is a classic compiler optimization that can eliminate and spread out loop carried dependencies, and thus has the potential to reduce LSU delay. Normally, unrolling would only be used when spin up and spin down costs are overcome by reducing the number of executed instructions. However, TS-aware compilation provides a new use for unrolling — avoiding errors to increase the efficiency of TS by grouping often independent instructions (like vector math) and eliminating often dependent instructions (like branches and index updates). Unrolling also allows optimization of register allocation over multiple loop iterations that can eliminate loads and stores, thus reducing pressure on the LSU. Unrolling can also reduce pressure on the branch resolution unit and arithmetic unit, since the number and frequency of branch instructions and loop index updates are reduced. Thus, in addition to fostering critical path avoidance by reducing dependencies, loop unrolling can also be an agent for activity throttling.

Figure 4.4 shows an example of loop unrolling. Figure 4.5 demonstrates that the compiling to avoid dependencies and activity on the forwarding paths with loop unrolling alters the activity distribution of the LSU and significantly reduces the error rate of the processor. Figure 4.5

```
for(i=0; i<N; i++)            for(i=0; i<N; i+=4){
    sum += A[i];                 sum1 += A[i];
                                 sum2 += A[i+1];
                                 sum3 += A[i+2];
                                 sum4 += A[i+3];
                             }
                             sum=sum1+sum2+sum3+sum4;
```

Fig. 4.4 Original loop (left) and unrolled loop (right).

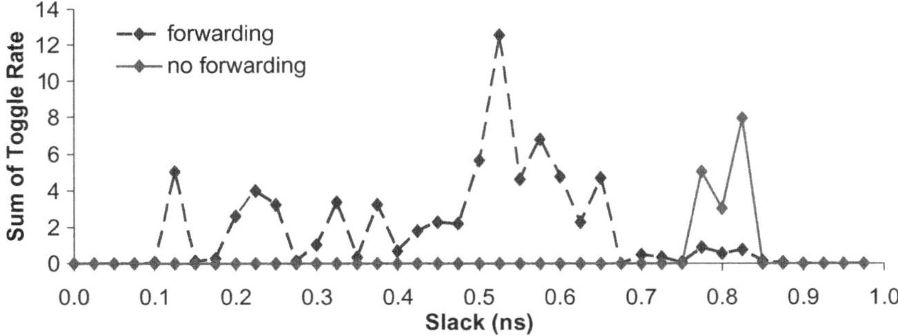

Fig. 4.5 Since forwarding paths are critical in the LSU, eliminating dependencies and the need for store-to-load forwarding reduces activity on the critical paths of the LSU.

shows the activity-weighted (dynamic) slack distribution of the LSU for the two instruction streams of Figure 4.4. The figure demonstrates that activity on the critical paths of the LSU is greatly reduced when loop unrolling is used. The dependencies in the original code are spread apart so that they do not simultaneously reside in the LSQ, and forwarding is not required.

Figure 4.6 shows the error rate of the processor when executing the two code sequences of Figure 4.4. Unrolling significantly reduces the error rate by reducing activity on the forwarding paths in the LSU. This error rate reduction enables additional overscaling and results in a substantial energy reduction for a Razor-based TS processor, as shown in Table 4.1.

In the normal, error-free case, the same unrolled loop provides a meager energy benefit of only 2%, on average, as unrolling causes dynamic power to increase significantly, even as it increases throughput. This further demonstrates the need for TS-aware compiler analysis and optimization.

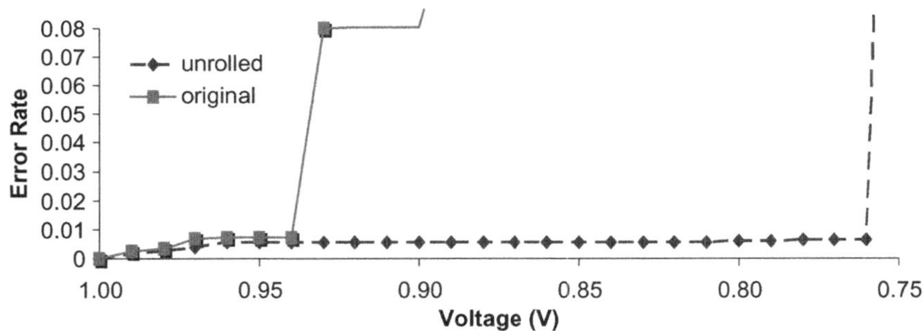

Fig. 4.6 Loop unrolling reduces activity on LSU forwarding paths, resulting in a significant error rate reduction.

Table 4.1.   Razor-based TS and error-free energy savings (%) for loop unrolling.

CORE	original	unrolled	unrolled error-free
ss1	11.8	43.1	1.6
ss2	6.4	20.8	2.0
ss4	4.0	42.9	3.2

Other compiler optimizations that target critical path avoidance and critical activity throttling have also been demonstrated to be effective in reducing the error rate and improving the energy efficiency of error resilient processors [36]. Balancing the amount of instruction-level parallelism exposed to the processor backend, and splitting or fusing loops have potential to manipulate the activity distribution of the processor and throttle or avoid activity on error prone paths. Additionally, TS-aware compilation work has demonstrated that gcc optimization flags can be manipulated to influence the energy efficiency of binaries on TS processors.

5

Scheduling and Mapping for Stochastic Processors

Work has been proposed to address the problem of task scheduling and mapping on heterogeneous multi-core processors [44]. This work is relevant to stochastic computing in that one axis of heterogeneity between cores may be reliability. The work seeks to achieve performance and energy efficiency by assessing the differences between tasks and between cores and finding the task-to-core mapping that optimizes a given efficiency metric.

The work proposes to use a graph-based program representation called System-level Instruction Set Architecture (SISA) to facilitate the task of mapping tasks to heterogeneous cores that typically have different ISAs. SISA represents programs as graphs with application characteristics such as data communication, length of computational tasks, reliability requirements, and task dependency. This allows for pre-running of tasks to identify the most efficient core on which to execute and dynamic task management to improve scheduling efficiency. SISA performs static scheduling based on Integer linear programming.

6

Application and Programming Language Support for Stochastic Processors

Stochastic processors enable reliability to be traded for increased energy efficiency when some form of error resilience is available. While applications from some classes often exhibit natural error tolerance (e.g., data-intensive applications), other classes of applications may not be naturally amenable to making energy-reliability tradeoffs. In this section, we discuss application and programing language techniques that aim to manage the impact of errors on a program or make applications more suitable for execution on stochastic processors that can allow errors to propagate from hardware to software.

EnerJ [32] is a programming language extension that gives programmers more control over how programs execute on stochastic processors. EnerJ specifies how a stochastic processor is allowed to make energy-reliability tradeoffs for a given piece of code by allowing the programmer to specify which variables in the program are allowed to be approximate and which should be precise. Variables that are marked as approximate can take advantage of energy-reliability tradeoffs, for example, by using unreliable memories and approximate or error-prone arithmetic units.

To maintain the dichotomy between approximate and precise program state, EnerJ performs type checking that prevents an approximate variable from affecting the assignment of a precise variable. If approximate variables need to be used in combination with precise variables, EnerJ allows an approximate variable to be "endorsed" so that it may be used in future precise computations.

Relax [13] is a framework for managing stochasticity in hardware by specifying how errors should be allowed to affect software. The potential benefits of Relax come from relaxing correctness guards imposed on hardware and allowing errors to propagate to software. The Relax framework uses an ISA extension, along with a try-catch-like software construct, that allows a programmer to specify regions of code in which correctness can be relaxed and errors can be allowed to propagate to software.

Relax relies on hardware error detection but does not support hardware error recovery, because of the relatively high cost of correcting errors in hardware. Instead, Relax performs error recovery in software. The Relax framework allows the programmer to specify an error rate for a given block of code. Relax treats all errors as equals, using a single, user-specified, recovery strategy such as retry or ignore when an error is detected. In the current formulation of Relax, faulty results are always discarded. Software frameworks like Relax may be useful for managing stochasticity in programs that run on programmable stochastic processors.

Application robustification [39] proposes an alternative approach to variation tolerance that exposes variations to software rather than detecting and correcting them in hardware. Application robustification work claims that it may be more energy efficient to make a program robust to errors and execute the program on low-power, stochastic hardware, rather than leaving the program alone and executing it on high-power hardware that is robust to errors. In order to ensure acceptable output quality on hardware that may make errors, application robustification proposes to transform applications into numerical optimization problems that can be solved with stochastic optimization techniques. Since these optimization techniques converge to the correct result, even when computations are noisy, the robustified applications are naturally error tolerant. Although the robust, stochastic optimization version of a program may take many iterations to converge to an acceptable result, each iteration of the program has a low energy cost. In this way, it is possible to save energy over a deterministic program execution. Furthermore, as variations become more common, application robustification may become useful simply as a means of achieving acceptable results on hardware that is necessarily stochastic [22, 39].

While some applications require application robustification techniques to achieve acceptable results on stochastic hardware, other applications exhibit natural error tolerance. Algorithmic approximate correction [37] relies on the inherent error tolerance exhibited by many applications and applies approximate error correction to improve output quality, resulting in a higher quality, though potentially noisy output. The goal is to ensure that even in the presence of errors, output quality is not degraded by more than an acceptable threshold. Algorithmic approximate correction can be used in different scenarios to provide performance or output quality guarantees for an application running on a stochastic processor.

7

Testing Techniques

Increased variability in semiconductor manufacturing not only necessitates significant shifts in hardware and software development, but also in the realm of chip testing. Due to the expected negative impact of variations on chip yields, new intelligible testing techniques [4, 5] have been proposed to maintain acceptable and profitable yields in spite of high defect rates.

Intelligible testing identifies and addresses two potentially limiting characteristics of conventional testing techniques. First, conventional testing techniques classify all chips on a pass/fail basis. There is no range of gradation between these two extremes. Second, conventional testing techniques do not take application characteristics into account and thus cannot rate chips based on how they function for specific target applications. Because classic testing strategies create a strict dichotomy between defect-free and faulty chips, a single hardware fault can cause a chip to be discarded, even if the chip could work acceptably for some applications, regardless of the fault. As we enter the multi-core era, it may become increasingly likely that chips contain circuitry that is not used for all applications. Consequently, scenarios in which faulty chips can function acceptably may also become more prevalent.

In order to determine which faulty chips may be salvaged, intelligible testing proposes a richer set of metrics to characterize how different faults translate into errors on a chip. In addition to the classical pass/fail dichotomy created by conventional testing techniques, intelligible testing techniques classify faulty chips based on three additional, "stochastic" metrics. *Error rate* conveys how frequently a fault manifests as an error. *Error significance* characterizes the impact that an error has on output quality or performance. Finally, *error accumulation* refers to the fact that some errors manifest once and vanish while others infect system state and continue to produce more errors and/or more significant errors in the future. With these metrics in mind, intelligible testing revisits the chips that fail initial defect testing to see whether their faults can be tolerated.

Some faults may not affect the behavior of all applications. For example, a defect in the FPU of a processor would likely only degrade the output quality and performance of floating point applications. For integer applications, degradation may be minimal or non-existent. Other faults may only degrade performance or output quality without causing application failures. If these degradations are tolerable, chips with such faults can still be used. For example, a defect in a cache may necessitate using only a subset of the original cache lines. While this will likely degrade performance, correctness is not affected.

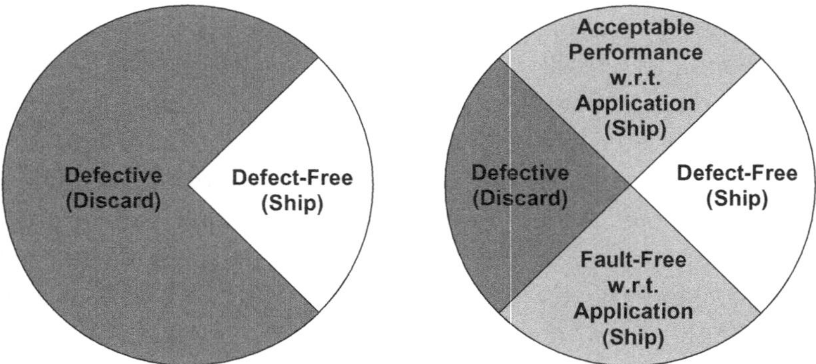

Fig. 7.1 Intelligible testing techniques increase yield by identifying and salvaging chips that have faults but are fault free or exhibit acceptable performance and output quality for certain target applications [5].

Intelligible testing increases yield by salvaging chips in these two categories (Figure 7.1 [5]). Although these chips have faults, they are either fault-free for some target applications, or they have acceptable performance and output quality for the target applications, despite the manifestation of occasional faults.

One limitation of intelligible testing is that it increases test time and cost. Normally, test time is restricted for economic reasons. However, the positive effect on yield due to salvaged chips may be worth the extra testing cost, especially as variability continues to increase with technology scaling.

8

Other Work

In this section, we discuss other works related to the field of stochastic computing.

Probabilistic CMOS (PCMOS) [8] attempts to exploit the stochasticity of low-energy circuits as a source of randomness in inherently probabilistic applications. PCMOS claims that an inverter operating close to the thermal noise margin will act as a probabilistic bit, with a tunable probability of outputting a "1" or a "0." Since operating near the thermal noise margin requires a much lower voltage than the nominal voltage, PCMOS logic can produce "random" outputs with low power consumption. Thus, an application that uses random values should be able to save energy by generating the random values with PCMOS logic.

PCMOS work proposes to create application-specific random value generators, aside from deterministic logic, such that there is a strict partitioning between probabilistic and deterministic components in a PCMOS design. Thus, the probabilistic logic acts as a co-processor that is polled whenever the deterministic components require random values. Because of the strict partitioning that is required, the PCMOS design style may only be suitable for algorithms with well-defined probabilistic steps. Strict partitioning also incurs costs for communication between the deterministic host and the probabilistic co-processor. If the probabilistic step is critical to an application, this communication link may become a bottleneck.

On another note, PCMOS design relies on high quality and controllability of random values produced by probabilistic logic, since the values are integral to application output quality. It should be noted that the efficiency of PCMOS logic depends on the desired output probability distribution. Also, supporting a wide range of probabilities degrades the efficiency of PCMOS.

Scalable Effort Hardware Design [9] exploits algorithmic error tolerance in order to improve the energy efficiency of hardware. Since some algorithms are naturally tolerant to errors, scalable effort hardware design proposes to relax the traditional requirement for exact equivalence between the hardware specification and the hardware implementation. A hardware design that approximately adheres to the design specifications may provide acceptable output quality when running a robust application. The focus of this work is to identify mechanisms at the circuit, architecture, and algorithm levels that influence the exactness or correctness of the final result, and expose these mechanisms as knobs during design optimization. In this way, output quality can be traded for energy efficiency.

Proponents of this work stress the importance of cross-layer optimizations, claiming that simultaneous consideration of optimizations at all design layers results in a more efficient design than when optimizations in each layer are considered separately. This claim should easily hold true, since simultaneous consideration of more axes of optimization should prevent locally optimal, globally sub-optimal decisions. Nevertheless, in order to maintain truly "scalable" effort in the hardware design, one should take care not to overly increase the complexity of making design decisions. Whenever one considers many layers of design optimization simultaneously, the optimization space that must be evaluated to make a single decision explodes, and the complexity of making each decision grows.

Code perforation [21] also takes advantage of noise-tolerance in applications to reduce energy. Proponents of code perforation argue that while some applications already trade accuracy for performance, the tradeoffs are typically application specific, as they require algorithmic changes. Instead, code perforation focuses on program modifications that can be made automatically by a compiler to trade accuracy for performance. The idea relies on the assumption that some programs can achieve acceptable output quality even if some of the operations in the program are forgone. As such, code perforation proposes to skip nonessential lines of code in order to increase performance. By monitoring the effect of code perforation, distortion is kept within user-defined bounds. Since both output quality and performance are monitored, the code perforation compiler can either maximize performance for a given output quality or maximize output quality for a given performance target. For many applications where code perforation can be applied, perforating code would be similar to changing the size of the sample population in a statistical sampling-based problem.

9

Summary and Future Work

Shrinking device sizes and growing static and dynamic nondeterminism challenge the reliable manufacturing and operation of circuits. Uncertainty in performance, power, and reliability results in decreased efficiency, since correctness must be ensured on inherently noisy hardware. The rigid correctness contract between hardware and software leaves potential performance and energy benefits untapped. As conventional design approaches that ensure determinism, even under worst case variations, become increasingly inefficient, a new paradigm for stochastic computing has arrived. Rather than hiding variations under expensive guardbands, stochastic designs relax traditional correctness constraints and deliberately expose hardware variability to higher levels of the compute stack, thus tapping into potentially significant performance and energy benefits, but also opening the potential for errors.

Stochastic computing techniques have been proposed at many levels of the computing stack, including design-level techniques that manipulate the error distribution of hardware, architecture frameworks that adapt to stochasticity, compiler optimizations that increase the efficiency of programmable stochastic processors, and testing techniques that increase yield with stochastic test criteria.

Going forward, as variability continues to grow, stochastic design optimizations, architecture frameworks, compiler optimizations, and testing techniques will be essential to achieving energy and performance efficiency. Though previous works have made inroads in these areas, the landscape is still ripe for exploration. For instance, stochastic design-level optimizations can benefit from approximate synthesis techniques that transform a high-level hardware description into the most efficient approximate version of the hardware. Stochastic architecture frameworks can improve efficiency by ceasing to treat all errors as equals and considering instead the software-level criticality of errors. Compiler-level work can take advantage of dynamic compilation techniques to route instructions around errors, and static compilation techniques that are more robust to errors. In testing, yield and profit could be increased with low-overhead, fine-grained binning strategies that deem more chips useful for specific purposes. As static and dynamic nondeterminism continue to increase, the potential of stochastic computing to reap significant benefits will grow as well.

References

[1] T. Austin, V. Bertacco, D. Blaauw, and T. Mudge, "Opportunities and challenges for better than worst-case design," in *Proceedings of ASPDAC*, pp. 2–7, 2005.

[2] K. Bernstein, D. Frank, A. Gattiker, W. Haensch, B. Ji, S. Nassif, E. Nowak, D. Pearson, and N. Rohrer, "High-performance CMOS variability in the 65-nm regime and beyond," *IBM Journal of Research and Development*, vol. 50, no. 4, pp. 433–449, 2006.

[3] K. Bowman, J. Tschanz, C. Wilkerson, S. Lu, T. Karnik, V. De, and S. Borkar, "Circuit techniques for dynamic variation tolerance," in *DAC*, pp. 4–7, 2009.

[4] M. Breuer, "Intelligible test techniques to support error-tolerance," in *ATS*, pp. 386–393, 2004.

[5] M. Breuer and S. Gupta, "Intelligible testing," in *MTAS*, 1999.

[6] T. Burd, S. Member, T. Pering, A. Stratakos, and R. Brodersen, "A dynamic voltage scaled microprocessor system," *IEEE Journal of Solid-State Circuits*, vol. 11, no. 35, pp. 1571–1580, 2000.

[7] Y. Cao, P. Gupta, A. Kahng, D. Sylvester, and J. Yang, "Design sensitivities to variability: Extrapolations and assessments in nanometer VLSI," in *Proceedings of IEEE ASIC/SOC Conference*, pp. 411–415, 2002.

[8] L. Chakrapani, B. Akgul, S. Cheemalavagu, P. Korkmaz, K. Palem, and B. Seshasayee, "Ultra-efficient (Embedded) SOC architectures based on probabilistic CMOS (PCMOS) technology," in *Proceedings of DATE*, pp. 1110–1115, 2006.

[9] V. Chippa, D. Mohapatra, A. Raghunathan, K. Roy, and S. Chakradhar, "Scalable effort hardware design: Exploiting algorithmic resilience for energy efficiency," in *DAC*, pp. 555–560, 2010.

[10] N. Choudhary, S. Wadhavkar, T. Shah, S. Navada, H. Najaf-abadi, and E. Rotenberg, "FabScalar," in *WARP*, 2009.

[11] J. Cong and K. Minkovich, "Logic synthesis for better than worst-case designs," in *VLSI-DAT*, pp. 166–169, 2009.

[12] S. Das, C. Tokunaga, S. Pant, W. Ma, S. Kalaiselvan, K. Lai, D. Bull, and D. Blaauw, "Razor II: In situ error detection and correction for PVT and SER tolerance," in *Proceedings of ISSCC*, pp. 400–622, 2008.

[13] M. de Kruijf, S. Nomura, and K. Sankaralingam, "Relax: An architectural framework for software recovery of hardware faults," in *ISCA*, 2010.

[14] S. Dhar, D. Maksimovic, and B. Kranzen, "Closed-loop adaptive voltage scaling controller for standard-cell ASICS," in *IEEE/ACM ISLPED*, pp. 103–107, 2002.

[15] D. Ernst, N. Kim, S. Das, S. Pant, R. Rao, T. Pham, C. Ziesler, D. Blaauw, T. Austin, K. Flautner, and T. Mudge, "Razor: A low-power pipeline based on circuit-level timing speculation," in *Proceedings of IEEE/ACM MICRO*, pp. 7–18, 2003.

[16] S. Ghosh, S. Bhunia, and K. Roy, "CRISTA: A new paradigm for low-power, variation-tolerant, and adaptive circuit synthesis using critical path isolation," *IEEE TCAD*, vol. 26, no. 11, pp. 1947–1956, 2007.

[17] B. Greskamp and J. Torrellas, "Paceline: Improving single-thread performance in nanoscale CMPs through core overclocking," in *PACT*, 2007.

[18] B. Greskamp, L. Wan, W. R. Karpuzcu, J. J. Cook, J. Torrellas, D. Chen, and C. Zilles, "BlueShift: Designing processors for timing speculation from the ground up," in *Proceedings of IEEE HPCA*, pp. 213–224, 2009.

[19] P. Gupta, "Design for ultra-low-k1 patterning and manufacturing," in *Tutorial at ICMTS*, 2009.

[20] R. Hegde and N. Shanbhag, "Energy-efficient signal processing via algorithmic noise-tolerance," in *ISLPED*, pp. 30–35, 1999.

[21] H. Hoffmann, S. Sidiroglou, M. Carbin, S. Misailovic, A. Agarwal, and M. Rinard, "Dynamic knobs for responsive power-aware computing," in *ASPLOS*, pp. 199–212, 2011.

[22] K. Huang and J. Abraham, "Algorithm-based fault tolerance for matrix operations," *IEEE Transactions on Computing*, vol. 33, no. 6, pp. 518–528, 1984.

[23] ITRS, "International technology roadmap for semiconductors," http://www.itrs.net/reports.html, 2010.

[24] S. Jones, "Exponential trends in the integrated circuit industry," http://www.icknowledge.com, 2008.

[25] A. Kahng, S. Kang, R. Kumar, and J. Sartori, "Designing processors from the ground up to allow voltage/reliability tradeoffs," in *IEEE HPCA*, 2010.

[26] A. Kahng, S. Kang, R. Kumar, and J. Sartori, "Recovery-driven design: A methodology for power minimization for error tolerant processor modules," in *ACM/IEEE DAC*, 2010.

[27] A. Kahng, S. Kang, R. Kumar, and J. Sartori, "Slack redistribution for graceful degradation under voltage overscaling," in *IEEE/SIGDA ASPDAC*, 2010.

[28] T. Kehl, "Hardware self-tuning and circuit performance monitoring," *IEEE International Conference on Computer Deisgn*, pp. 188–192, 1993.

[29] L. Leem, H. Cho, J. Bau, Q. Jacobson, and S. Mitra, "ERSA: Error resilient system architecture for probabilistic applications," 2010.

[30] S. Palacharla, N. Jouppi, and J. Smith, "Complexity-effective superscalar processors," in *ISCA*, 1997.

[31] Y. Pan, J. Kong, S. Ozdemir, G. Memik, and S. Chung, "Selective wordline voltage boosting for caches to manage yield under process variations," in *DAC*, pp. 57–62, 2009.

[32] A. Sampson, W. Dietl, E. Fortuna, D. Gnanapragasam, L. Ceze, and D. Grossman, "EnerJ: Approximate data types for safe and general low-power computation," in *PLDI*, 2011.

[33] S. Sarangi, B. Greskamp, A. Tiwari, and J. Torrellas, "EVAL: Utilizing processors with variation-induced timing errors," in *IEEE/ACM MICRO*, pp. 423–434, 2008.

[34] J. Sartori and R. Kumar, "Overscaling-friendly timing speculation architectures," in *ACM/IEEE GLSVLSI*, 2010.

[35] J. Sartori and R. Kumar, "Architecting processors to allow voltage/reliability tradeoffs," in *CASES*, 2011.

[36] J. Sartori and R. Kumar, "Compiling for timing error resilient processors," in *TECHCON*, 2011.

[37] J. Sartori, J. Sloan, and R. Kumar, "Stochastic computing: Embracing errors in architecture and design of processors and applications," in *CASES*, 2011.

[38] N. Shanbhag, R. Abdallah, R. Kumar, and D. Jones, "Stochastic computation," in *Proceedings of DAC*, pp. 859–864, 2010.

[39] J. Sloan, D. Kesler, R. Kumar, and A. Rahimi, "A numerical optimization-based methodology for application robustification: Transforming applications for error tolerance," in *DSN*, 2010.

[40] Sun, "Sun OpenSPARC project," http://www.sun.com/processors/opensparc/, 2010.

[41] University of Michigan, "Bug Underground," http://bug.eecs.umich.edu/, 2007.

[42] G. Varatkar, S. Narayanan, N. Shanbhag, and D. Jones, "Variation-tolerant, low-power PN-code acquisition using stochastic Sensor NOC," 2008.

[43] L. Wan and D. Chen, "DynaTune: Circuit-level optimization for timing speculation considering dynamic path behavior," in *ICCAD*, 2009.

[44] Y. Yetim, W. Jia, S. Malik, M. Martonosi, and K. Shaw, "A system-level ISA and its applications to energy-performance-reliability scheduling and scratchpad allocation," http://www.gigascale.org/pubs/2341.html, 2010.

Lightning Source UK Ltd.
Milton Keynes UK
UKOW07f2333111214

242988UK00002B/109/P